Fresh Clean Home

Wendy Graham

Fresh Clean Home

MAKE YOUR OWN NATURAL CLEANING PRODUCTS

PAVILION

First published in the United Kingdom in 2018 by
Pavilion
43 Great Ormond Street
London
WC1N 3HZ

Distributed in the United States and Canada by
Sterling Publishing Co, Inc.
1166 Avenue of the Americas
New York, NY 10036

ISBN 978-1-911595-10-6

A CIP catalogue record for this book is available from the British Library.

10 9 8 7 6 5 4 3 2 1

The information and material provided in this publication is representative
of the author's opinions and views. The information and material is
presented in good faith; however, no warranty is given, nor are results
guaranteed. Pavilion does not have any control over, or any responsibility
for, any author or third party websites referred to in this book.

Reproduction by Mission Productions, Hong Kong
Printed and bound by 1010 Printing International Ltd, China

This book can be ordered direct from the publisher at
www.pavilionbooks.com

AUTHOR'S DISCLAIMER

The recipes presented in this book are those that – among others –
the author uses in her own home. They have not been scientifically
tested, and neither the author nor the Publisher can guarantee that
they will produce the intended results. Readers follow and use recipes
at their own risk. Always patch test a recipe on an inconspicuous area,
and always follow the use and care instructions both in the recipes
themselves and on the packaging for any ingredients. If in doubt you
should seek advice from a specialist or the manufacturer.

PUBLISHER'S NOTES

Essential oils can be potent. Always read the labels before using,
check for contraindications and ensure safe dilutions in a base oil
or soap. Note that many of the recipes use citrus fruit or oils. Citrus
juice can damage natural stone surfaces, as can other highly acidic
products such as vinegar. Please read the introduction, including
safety instructions on page 20, before you begin. Never mix hydrogen
peroxide and vinegar (see page 22). Neither the author nor the
publisher can take any responsibility for any illness, injury or damage
caused as a result of following any of the advice or using any of the
recipes or methods contained in this book.

Recipe Menu

Introduction

From reducing the number of harsh chemicals you use in your home to saving money, or living a little lighter on the Earth (and everything in between), there are myriad reasons to want to try making your own natural cleaning products.

I have been making my own cleaning products for more than a decade, and have amassed a host of natural and effective recipes to clean every corner of my home.

So how did it all begin? Over ten years ago, I had some laundry that kept coming out of the washing machine with a funny smell. It was exasperating! I couldn't figure out the problem, until – after some rather extensive research – I discovered that the fabric conditioner I'd been buying merrily for years was to blame. It had gummed up our washing machine, making it a breeding ground for bacteria. Not only that: I also discovered that the way that conventional fabric conditioner works is to cover your clothes with a waxy coating that softens the fabric, but it makes the fabric less able to absorb water and detergent, which locks in bad odours. Needless to say the first recipe I ever made was natural fabric conditioner, after giving my machine a thorough clean.

Although I was exceptionally dubious to begin with, I was surprised to find that the natural methods I experimented with solved our machine issue completely. It was a lightbulb moment for me: I now knew that natural methods could work more effectively than conventional products, and be so much cheaper. That was it. I was on a mission, slowly but surely, to replace the cleaning products I use in my home with the homemade versions I'm about to share with you.

If you've tried making natural cleaning products before and have been disappointed with the results, I want to encourage you to try again – there could be good reasons why your first attempts didn't work. After all, this is delicate chemistry. And if you think natural cleaning is all about vinegar, think again. While I think vinegar definitely has a place in natural cleaning, I don't think it's the most effective cleaner for every situation – you'll find that many of my recipes are vinegar-free.

How to Begin

Start slowly. Use up the products you already have and as they run out try making their natural replacements (don't throw away any pump sprays or bottles – give them a thorough clean-out and save them to fill with your own homemade products). Also, remember that this isn't a case of all or nothing – if you find there are certain shop-bought products you can't do without, that's okay. Even I have a few staples that to this day find their way into my shopping trolley. I hold my hands up, and confess:

Washing-up liquid

I've found that an effective homemade washing-up liquid is too tricky – it requires specialist ingredients (such as anionic surfactants that lather and clean) and I find it's much easier and more cost effective to buy a bottle of eco-friendlier product. I not only wash up with it, but you'll see that I use it in several of my homemade recipes for other products, too.

Dishwasher tablets

Pretty much all commercial dishwasher tablets contain an ingredient called sodium silicate, which rinses away food and soap deposits, and is completely soluble in water. As a result your dishes always come out clean and streak-free. I've found that any homemade formula that doesn't contain sodium silicate just does not work – but sodium silicate is available only commercially. I concede: eco-friendlier shop-bought dishwasher tablets it is.

In short, don't get too bogged down in trying to replace everything. The more you make, the better for you, your home and the environment, but if that's only one or two products overall, that's still enough to make a positive difference. With that in mind let's get started with some of the basics behind natural cleaning – from what you'll need to a few notes on safety and effectiveness.

The Must-haves

I like to keep things as simple and as straightforward as possible, so I use a core group of easily sourced ingredients to make my cleaning products. Some of these you may already have in your cleaning armoury, or even lurking in your food cupboards.

Bicarbonate of soda

Not to be confused with baking powder, bicarbonate of soda (also known as baking soda) is a natural cleaning staple (it just happens to be used in baking, too). Rather than buying the little tubs from the supermarket, look for 500g/17oz boxes that are often sold in hardware shops and discount stores. Alternatively, you can buy in bulk online. It's the same stuff you would use in baking, but doesn't come in food-safe packaging, so is a lot cheaper.

Glycerine

A natural thickener made from plant oils and often used in baking, glycerine is brilliant when you're making something that needs a thicker consistency, such as the oven-cleaning gel on page 44. It also helps to emulsify essential oils with water. You can buy glycerine quite cheaply from the baking aisle in any supermarket. I use Dr Oetker as it's vegetarian friendly.

Borax substitute

Also known as sodium sesquicarbonate, borax substitute is chemically very similar to borax (sodium borate, from the mineral boron), which was banned in the UK and EU in 2010 as a result of safety scares. The substitute, which is believed to carry none of the harmful effects of boron, comes from mineral deposits that have brilliant laundry cleaning and general cleaning properties, especially in the bathroom. It's available in hardware and discount stores, and costs little more than loose change for a 500g/17oz box. If you buy online, you'll bring the cost down even further.

Hydrogen peroxide

Did you think only bleach could disinfect? Let me open your eyes to the wonders of hydrogen peroxide.

The chemical symbol of hydrogen peroxide is H_2O_2 – basically, that's water with one extra oxygen atom. Put like that hydrogen peroxide might seem innocuous, but in fact it's bad news for germs and bacteria. That extra oxygen atom is highly volatile and causes oxidation, a reaction in which the hydrogen peroxide steals electrons from bacteria, breaking down their cell walls. With no cell walls, bacteria, which are simple micro-organisms, die, leaving surfaces germ free.

In addition, it doesn't leave any harsh chemical residue on your natural surfaces, because when hydrogen peroxide reacts with organic material it breaks down into oxygen and water. All this makes hydrogen peroxide an extremely effective disinfectant without the cloying odour or eye- and lung-irritating properties of bleach.

Hydrogen peroxide is commonly used as a mild skin antiseptic and mouthwash, but its sale is a little restricted, so it's sold only over-the-counter by pharmacists. You can buy it in different concentrations – I buy the 3% stuff.

Jojoba oil

Jojoba oil isn't actually an oil, but a liquid wax extracted from the seed of the jojoba plant. It has an exceptionally long shelf life (unlike olive oil it doesn't go rancid quickly), making one small bottle a very economical purchase. Shop around online to pick it up cheaply.

Citric acid

A highly concentrated fruit acid, citric acid is commonly used as a preservative in baking and brewing, but it is also a brilliant cleaner, particularly when it comes to tackling calcium deposits in hard-water areas. Look for it in homebrew shops, Asian supermarkets, or online. As citric acid is highly concentrated, take particular care when mixing it in the recipes – always wear gloves and eye protection, and work in a ventilated area.

Liquid castile soap

This simple liquid soap is made from pure plant oils, including olive and coconut oils, and is free from any artificial agents. It's completely biodegradable and has a host of natural cleaning applications.

I use Dr Bronner's castile soap and am a particular fan of the scented varieties. It can be a bit pricey to buy, so only ever use it diluted to get more for your money. Dr Bronner's is usually readily available in health-food stores, but you can buy unbranded, unscented liquid castile soap online a little more cheaply.

Witch hazel

With similar properties to vodka (see opposite), witch hazel makes a cheap and effective cleaner that's available from any pharmacy, or online.

Soda crystals

Also known as washing soda, soda crystals are similar to bicarbonate of soda but have been processed to give a much higher alkalinity that makes them tough on dirt and grease. They are great for cleaning your laundry and your kitchen! A bag of soda crystals will come cheap from most supermarkets (look in the cleaning aisle), or hardware or discount stores. Alternatively, buy in bulk online. Always wear gloves when working with soda crystals.

Vinegar

The staple in many green cleaning products, vinegar has its reputation as the king of natural cleaning for good reason – it's cheap, non-toxic, and cuts through dirt with ease. However, vinegar does have a particularly pungent odour that people either love or hate. Don't despair if you're a hater: once vinegar dries it's odourless.

Make sure you use only white vinegar to avoid staining. I buy in bulk from eBay, but you can also try Asian supermarkets, or buy smaller glass bottles of the stuff at the supermarket.

Vodka

I buy the cheapest own-brand straight vodka I can lay my hands on and use it specifically for cleaning. It's a great deodoriser, degreaser and also has some disinfectant properties. And, in case you're worried, no it doesn't leave a smell of alcohol behind. The recipes use only a little at a time, so you'll find one bottle lasts a long time.

Xanthan gum

A staple for any gluten-free baker, xanthan gum is made by fermenting corn sugar with the bacteria *Xanthomonas campestris* (the same bacteria that creates black spots on broccoli and cauliflower). It's sold as a powder and is used as a natural thickener for some cleaning products. It can be a little temperamental to work with (it doesn't dissolve in water), but follow the instructions I've given closely and you should be fine.

Essential Oils

Manufacturers don't have to list the ingredients that go into making up the fragrances of their cleaning products. They are allowed to regard them as trade secrets, simply stating 'perfume' on the label. But the word perfume can hide a reality of hundreds of different chemicals and petrochemicals.

Instead of synthetic perfumes I scent my products with pure essential oils, which are distilled directly from plants. In some recipes the oils are there for scent purposes only, but in most they also provide antibacterial and disinfectant properties. Here are some of my favourite oils and their properties, although feel free to experiment with any that you particularly like.

Bergamot (Citrus bergamia)

Antibiotic, antiseptic and disinfectant; air-freshening.

Eucalyptus (Eucalyptus globulus or Eucalyptus radiata)

Antibacterial, antifungal, antimicrobial and antiviral; commonly found in mouthwashes, toothpastes and other dental-hygiene products.

Geranium (Pelargonium graveolens)

Flower-scented with air-freshening properties.

Ginger (Zingiber officinale)

Antiseptic and antibacterial with a zingy scent.

Grapefruit (Citrus paradisi)

Degreasing and air-freshening.

Lavender (Lavandula angustifolia)

Antibacterial, antifungal and antiviral with a pleasing, popular scent.

Lemon (Citrus limon)

Powerhouse degreaser; antimicrobial, antifungal and antibacterial.

Peppermint (Mentha piperita)

Antibacterial and antiviral with a fresh scent.

Pine (Pinus sylvestris)

Antifungal, antimicrobial and antiviral with a fresh forest scent.

Rosemary (Rosmarinus officinalis)

Antifungal and antibacterial with a fresh herb scent.

Sweet Orange (Citrus aurantium dulcis)

Excellent degreaser; air-freshening.

Tea tree (Melaleuca alternifolia)

Powerhouse germ-fighter; antibacterial, antiviral and antimicrobial.

Note that oils may come with specific warnings about usage. Tea tree, for example, is reported as being toxic to cats; while eucalyptus and peppermint are not recommended for use around babies and children; and you should avoid rosemary if you are pregnant.

All my recipes call for high dilutions of the oils and I am confident that they are safe to use in cleaning products around my home. However, I encourage you to research each oil you intend to use to make sure it doesn't have any contraindications that could harm you or members of your household – always make your own mind up about what's safe for you. Always store oils out of reach of children and animals.

If you have very sensitive skin, you may find citrus-based oils too harsh when used in cleaning products that will come in repeated contact with your skin, such as the hand soap, laundry powder and fabric conditioner recipes. In this instance feel free to swap the citrus oils for any others of your choosing (a combination of lavender and thyme is a favourite for me) or omit the oils altogether.

If you need to save money, I recommend buying just two oils: lavender and tea tree. These are probably the most multi-purpose of all.

Equipment

The beauty of making your own cleaning products is that you don't need much in the way of specialist tools. The following are the few bits of equipment I keep to hand in my kitchen (and some of them I borrow from my baking cupboard). You'll probably already have lots of them yourself.

Blender food processor

My blender comes out only when I'm baking or making laundry powder. If, like me, you're going to use yours for dual purpose, be sure to give it a thorough wash each time.

Funnel

Any old funnel will do, but it's essential for getting liquid products into storage bottles without any wastage. I have a silicone funnel that stores flat that I particularly love.

Glass bottles & jars

I am a hoarder of glass bottles and jars – I can always find a use for them! The glass bottles that white vinegar is sold in are particularly useful, and rather serendipitously standard spray nozzles (which you can buy online separately) fit them perfectly. Kilner jars and jars with screwtop lids will be invaluable. Wash and store any that you can, or buy online or in low-cost household stores.

Sunlight can impact the effectiveness of essential oils, so I have some amber glass bottles for the cleaning products I store on my countertop. If you intend to keep your products in a cupboard out of sunlight, you don't necessarily need dark glass, apart from when you make something containing hydrogen peroxide. In this case always use amber glass. There are lots of stockists online.

Hand blender

I use a hand blender for recipes that include xanthan gum, because I've found blending the mixture creates a better gel than mixing by hand. Still, if you don't mind a bit of elbow grease, you could do without this particular bit of kit.

Kitchen scales

A cheap set of kitchen scales will suffice. One of the things about making cleaning products is that you don't need to be too exact with measurements – a little more or less of one ingredient isn't going to make too much difference. Once you've made the recipes a few times, you may be able to eyeball the ingredients you use.

Labels

It's important to label your products so as you know what they are, and the date when you made them. Washi tape is really handy, or you could use adhesive labels. I use an old-school, hand-operated Dymo labeller (cheap to buy online) because the embossed labels will survive getting wet better than handwritten, sticky versions.

Spray nozzles

Reuse the nozzles from any existing cleaning products you have, but give them a good wash first to ensure you don't mix the old, synthetic solution and your special homemade brew. If you need to buy nozzles, though, there are several inexpensive online stockists.

Is Natural Cleaning Chemical-free Cleaning?

It's a common misconception that natural cleaning is chemical free. Even though this book uses natural products, all those products are chemicals. Vinegar's chemical name, for example, is acetic acid and its chemical formula is CH_3COOH; borax substitute is sodium sesquicarbonate and its chemical formula is $Na_3H(CO_3)_2$. Without being over the top, remember that even salt and water are chemicals ($NaCl$ and H_2O, respectively).

So, although these are recipes for 'natural' products, it's still important to treat them with respect, using safety precautions as you would with shop-bought alternatives.

Safety first

Even though you are making your products yourself, there are still certain safety rules that you should follow to ensure you are always protecting yourself, your home and those who live in it.

- Always wear rubber gloves when you clean.
- Always store your products in labelled jars and bottles, giving the name of the product and the date you made it.
- Keep the products and raw ingredients out of the reach of children and pets.
- Never mix natural cleaning products with conventional products, such as bleach. Vinegar, for example, when mixed with bleach releases chlorine gas, which even at low levels can cause burning, watering eyes, breathing problems, and coughing.
- Always patch test before using a new homemade product freely.

Natural Cleaning No-nos

There are several ingredients you should never mix. This might be because the combinations are hazardous to health, or because you'll end up with a completely ineffective cleaning product. The chances are if you've tried natural cleaning before and been disappointed with the results, it's because you mixed two of the following ingredients together:

Hydrogen peroxide & vinegar

A WARNING!

Mixing hydrogen peroxide with vinegar releases a noxious gas called peracetic acid, which, if breathed in, can permanently damage your lungs, cause asthma, and irritate the skin and eyes.

Liquid castile soap & vinegar

Although harmless, mixing liquid castile soap and vinegar is essentially a waste of money and soap. Vinegar is an acid and the liquid castile soap is a base. When you combine an acid and a base, the two ingredients react to cancel each other out.

When I want to mix soap with vinegar, I prefer to use an eco-friendlier washing-up liquid, which I think performs better.

Bicarbonate of soda & vinegar

You may have heard that bicarbonate of soda and vinegar make a great foaming cleaner. I'm afraid to tell you that this isn't the case. Bicarbonate of soda is a base and, as you already know, vinegar is an acid. They cancel each other out, leaving you with nothing more than a salty water solution. Save this combination for unblocking drains: the foaming action can help dislodge debris, but nothing more.

Borax substitute or soda crystals & vinegar

Both borax substitute and soda crystals are bases, so if you were to mix either of them with vinegar, the two would cancel each other out.

A Word on Shelf Life

For each recipe I've specified an approximate shelf life, but please do use your own judgement. If anything smells or looks funny, then it's probably time to throw it away and remake it.

In many cases, particularly in recipes that include water, the shelf life is short – that's because water (even boiled water) harbours bacteria. I don't recommend using water straight from the tap, which can further shorten the shelf life.

Easy as One, Two, Three

Every recipe in the book has a series of drops to indicate how easy it is to put together. Truth is, all of the recipes are easy, so the drops are relative to one another – from ridiculously straightforward to the few that are just a teeny bit more involved.

 = **Easy**
As easy as a shake or a stir

 = **Medium**
A little more prep required

 = **Advanced**
Needs a bit of commitment

Now that you know practically all there is to know about natural cleaning, let's get on to making our cleaning products! I hope you enjoy the recipes!

KITCHEN
#heartofthehome

Cleaning your kitchen naturally is a breeze – a beautifully scented breeze at that. From delicious-smelling essential oils, fresh herbs straight from the garden, and ripe and juicy citrus fruits, there's no need for artificial fragrance here. Got some lemons? Great, you can clean your microwave, or whip up some kitchen cleaner. Got a grapefruit? Then you've got the perfect sink cleaner.

Of course, as lovely as it would be, you can't clean your kitchen with fruit and herbs alone: so the recipes contain some of the hardest working natural cleaning products known to humankind. From soda crystals to salt – your kitchen (even your oven!) will be sparkling clean in no time.

Some of my kitchen sprays contain vinegar, but if your worktops are made of granite, marble or stone, or someone in your household is sensitive to the smell of vinegar, then I've even provided alternatives.

KITCHEN

Kitchen Spray Your Way

I'm a big fan of the lemon, lavender and thyme cleaning concentrate on page 30 diluted as a kitchen spray, but if you don't want to wait for all those scents to infuse, and would rather whip up something straightaway with vinegar and essential oils, then here are some of my recommendations for a super-smelling kitchen spray – your way.

Makes: 500ml/17fl oz
Shelf life: up to 8 weeks

250ml/9fl oz white spirit vinegar
250ml/9fl oz cooled boiled water
¼ teaspoon glycerine
Choose one of the following
    essential oil blends:
- 12 drops cinnamon, 12 drops clove and 16 drops sweet orange
- 10 drops tea tree, 14 drops lemon and 16 drops grapefruit
- 12 drops peppermint, 12 drops lavender and 16 drops lemon
- 12 drops tea tree, 12 drops lemongrass and 16 drops lemon
- 12 drops peppermint, 12 drops rosemary and 12 drops grapefruit

500ml/17fl oz glass spray bottle

Put the vinegar, water and glycerine into the spray bottle, pop on the pump lid and give the bottle a shake to mix thoroughly.

Now for the fun part: choose one of my favourite essential oil blends from the list on the left, and add it to the base mixture to make your kitchen spray smell amazing.

To use
Shake well, then spray onto surfaces and wipe away. If you're cleaning wooden surfaces, always spray onto the cloth rather than directly onto the wood.

⚠ Do not use on granite, marble or other natural stone surfaces.

#ortrythis Add a total of 40 drops of your favourite essential oil(s) for a customised scent.

Lemon, Lavender & Thyme Concentrate

Makes: 500ml
Shelf life: indefinitely when
    concentrated; or about
    8 weeks once diluted

Approximately 500ml/17fl oz
    white vinegar
Peel of approximately
    8–10 lemons
Generous handful of thyme sprigs
Generous handful of lavender
    sprigs

1 litre/35fl oz lidded glass jar

Some people love the smell of vinegar – if that's you… you lucky thing, because white vinegar is a fantastic kitchen cleaner. It's cheap, cuts through grease like nothing else, and deodorises. Some people even swear by its use as a disinfectant.

If you don't like the smell, though, all is not lost. There are two ways to lessen its pungent aroma. The first is to add essential oils to the mixture; the second is a bit more thrifty and requires using fruit peelings and whatever herbs you have to hand.

One of my favourite combinations is lemon, lavender and thyme. To reduce waste, freeze lemon peelings after cooking or baking until you have gathered enough to make the concentrate. And if you can grow lavender and thyme in your garden or window box to be able to clip a few sprigs here and there, even better!

Add the vinegar, lemon peel and herbs to your jar. (If you're using frozen lemon peel there's no need to defrost it.) Then, add the vinegar, covering the lemon peel and herbs.

Give the contents a stir, then put on the lid and seal tight.

Leave the jar in a cool, dark place for about 14 days, or longer if you want a stronger scent. Then, strain the liquid through a sieve, pressing down on the lemon peel and herbs to squeeze as much liquid out of them as possible. Pour the strained liquid back into the jar, and pop the peel and herbs in your compost bin.

To use
You can use this concentrate whenever any cleaning recipe calls for vinegar. Alternatively, use it as an all-purpose kitchen-cleaning spray: half fill a spray bottle with the scented vinegar mixture and top up the other half with cooled boiled water (for a 50/50 dilution). Diluting is important because the concentrated citrus oils, undiluted, could stain lighter surfaces.

⚠ Do not use on marble, granite or other natural stone surfaces.

#ortrythis As alternatives to the lemon, thyme and lavender, you can try orange, grapefruit or lime skins; and herbs such as rosemary or oregano.

Vinegar-free Lavender & Mint Kitchen Spray

If vinegar isn't your thing, don't worry – natural cleaning products are still within reach. For example, this cleaning spray will make light work of grime to give your kitchen a good, deep clean and there's no vinegar in it at all. It smells pretty amazing, too – even the most nose-sensitive family members will be happy!

Makes: 500ml/17fl oz
Shelf life: up to 8 weeks

500ml/17fl oz just-boiled water
1 teaspoon soda crystals
2 tablespoons liquid castile soap
15 drops lavender essential oil
15 drops lemon essential oil
10 drops peppermint essential oil
¼ teaspoon glycerine

500ml/17fl oz glass spray bottle

Pour the freshly boiled water into a measuring jug, then add the soda crystals and stir until fully dissolved. Fill your spray bottle with the water-and-soda solution, then add the remaining ingredients. Pop on the pump lid and shake well to combine.

To use
Shake well, then spray onto surfaces and wipe away. If you're cleaning wooden surfaces, always spray onto the cloth rather than directly onto the wood.

⚠ Omit the lemon essential oil if using on porous stone surfaces.

Lavender & Tea Tree Stonesafe Kitchen Spray

You may have noticed I'm a big fan of using mixtures of vinegar and citrus-based oils when I'm cleaning. However, they can be damaging to granite, marble and other natural stone surfaces. As an alternative I've come up with this gentle spray, which will give your porous work surfaces a lovely streak-free shine without the hefty price tag of shop-bought specialist granite or marble cleaners.

Makes: 500ml/17fl oz
Shelf life: up to 8 weeks

450ml/16fl oz cooled boiled water
2 tablespoons vodka
2 teaspoons liquid castile soap
15 drops lavender essential oil
10 drops tea tree essential oil

500ml/17fl oz glass spray bottle

Put all the ingredients in the spray bottle, secure the pump lid and shake well to combine.

To use
Shake, then spray directly onto your worktop. Buff up with a clean, dry cloth.

#ortrythis If you fancy a different scent, try peppermint and tea tree or eucalyptus oil, or feel free to experiment with your own favourite non-citrus combinations.

Natural Disinfectants

One of the most common questions I get about natural cleaning is how do I tackle germs and bacteria?

I'm of the mindset that not all bacteria is bad bacteria, and that wiping out all the bacteria in your home with sprays that kill 99.9% of bacteria probably isn't the best for your immune system. Therefore, I believe that cleaning surfaces with the natural cleaning spray of your choice is more than adequate at keeping your home clean and healthy.

However, I also understand that sometimes you need a surface not just to be clean but to be really clean. Perhaps when you've been chopping raw meat, or maybe a pet or a child has an accident, then you want to spray something for that extra peace of mind. In times like this, I reach for the hydrogen peroxide.

I've written in detail all about the wonders of hydrogen peroxide on page 11, and you'll see it pops up in a few of my recipes, but it's also handy to keep a bottle around to use it neat. It's available cheaply over the counter from any chemist in varying percentages – I buy the 3% hydrogen peroxide, and use it as follows:

Disinfectant Spray

To use hydrogen peroxide as a disinfectant spray simply decant the hydrogen peroxide into an amber glass spray bottle and add the nozzle. Spray wherever you need to disinfect – for example, on. worktops or toilets. If you want to use it on soft furnishings do a spot test first, as the fabric may not be colour fast. Hydrogen peroxide loses its potency once opened, so stored in a spray bottle, it will last for one month.

For chopping boards that have been used to prepare raw meat, I adopt a slightly different method. A scientific study carried out in 1997 found that spraying vinegar and then hydrogen peroxide was effective in killing food-borne micro-organisms, such as E-coli, salmonella and listeria.

Therefore, first I wipe down the chopping board with hot soapy water. Then I spray undiluted vinegar on the board and wipe with a clean, damp cloth. Then I rinse the cloth, spray 3% undiluted hydrogen peroxide on the board and wipe again. Please don't mix the vinegar and hydrogen peroxide together (see the safety warning on page 22). However, I've found spraying vinegar, wiping and then spraying hydrogen peroxide is perfectly fine. Just remember to wipe between sprays to avoid mixing the two products!

Disinfecting Dishcloths

Disinfect dishcloths and sponges by placing the cloths and sponges in a bowl, and pouring a 50/50 solution of hydrogen peroxide and boiled water over them. Leave to soak for 30 minutes then remove from the solution, rinse and allow to dry.

⚠ Do not mix with vinegar.

⚠ Hydrogen peroxide has a high acidity, so do not use on natural stone surfaces, such as granite or marble.

Citrus Scouring Scrub

Makes: 90g/3¼oz
Shelf life: about 8 weeks

1 grapefruit, or 2 oranges
   or lemons
2 tablespoons bicarbonate of soda
3 tablespoons coarse salt

100g/3½oz lidded jar

If you have a ceramic kitchen sink, you'll know how much cleaning it needs to stay grime-free – especially when it comes to the remnants of wheat-biscuit cereals, which in our house left me with scrubbing I used to dread. Some time ago I found a version of this scouring scrub on a great website called Crunchy Betty and I adapted it to suit my own ceramic-sink/caked-on-cereal needs.

The star of the show is the citrus peel. Gently dried and then finely ground, it's packed full of fruit oils that, combined with the gentle abrasiveness of the bicarbonate of soda and the salt, make light work of dirt and grime. Beautiful smelling, this scrub will leave your sink with a delightful citrus zing. It's also a fantastic way to use up citrus peel that you'd otherwise bin. This scrub does take a little bit of patience to make, but good things come to those who wait.

Use a paring knife to remove the peel from your chosen citrus fruit. Tear the peel into pieces, each no bigger than the size of a 10-pence piece, and put the pieces in a dish. Place the dish in a warm, dry spot. Leave the peel to dry out, turning it in the dish at least once a day, until it is hard. This could take about 4 days.

Place the dried peel in a food processor or blender and blitz for a few seconds until you have a soft, fine powder.

Combine the powdered peel with the bicarbonate of soda and salt, place in a clean, dry jar and secure the lid.

To use

Sprinkle the scouring powder liberally onto the surface you want to clean. Using a damp cloth, scrub the area, adding more powder as required, then rinse off.

#ortrythis This scouring powder is ideal on ceramic sinks and toilets, but may be too abrasive for acrylic baths or shower trays. If you want something gentler try my Supersoft Citrus Shower & Bath Scrub on page 80.

Quick & Easy Grapefruit Sink Cleaner

Makes: single use
Shelf life: use immediately

½ grapefruit
About 3 tablespoons coarse salt,
 plus extra if needed

If you're in a bit of hurry and have a grapefruit to hand, try this simple citrus sink cleaner. I swear cleaning your sink never smelled so good!

Using just two ingredients – grapefruit and salt – it's a really effective way to clean. The grapefruit juice degreases, while the salt acts as an abrasive to remove any dirt. If you're out of grapefruit, any citrus fruit will do. In fact, this grime-buster is a perfect way to use up any citrus fruit that might be past its prime.

Sprinkle the grapefruit half with a little coarse salt. Rinse out your sink and leave it wet, then sprinkle a large spoonful of salt in it.

To use

Take your salted grapefruit and get scrubbing. Every so often pausing to squeeze the grapefruit to release some juice, and to pick up more salt from around the sink.

Once the sink is clean, rinse it down again and wash the pulpy bits of grapefruit flesh down the drain. Particularly stubbornly stained sinks may require several halves of grapefruit to restore them to their former glory.

#ortrythis I recommend using this scrub on metal or porcelain sinks only, as the salt may be too abrasive for those made of acrylic. In that case use my Supersoft Citrus Shower & Bath Scrub on page 80.

* This scrub is also great for tackling rust stains on a metal sink or bath.
* Remember to save the peel to make the cleaning concentrate on page 30.

Long-lasting Fridge Freshener

A stinky fridge happens to the best of us. Take your eye off a vegetable past its prime or spill some milk and you've got a stink to high heaven. Thankfully, though, this is an easy one to tackle. All you need to do is turn to your old friends – elbow grease and bicarbonate of soda – to give that stench the heave-ho.

Makes: single use
Shelf life: up to 3 months
(in the fridge)

3 tablespoons bicarbonate of soda
5 drops lemon essential oil
(optional)

Small bowl or tub, or a small wide-
necked jar

Put the bicarbonate of soda in a small bowl, tub or jar and add the essential oil, if using. Stir to combine.

To use
Remove all the food from your fridge and give the inside a thorough clean (using your cleaning spray of choice), then put back the food (discard any that is past its best).

Place the bowl, tub or jar of bicarbonate of soda with the lemon essential oil, if using, on the middle shelf of the fridge and let it get to work. It will continue to absorb bad odours for about 3 months.

Clean & Green Oven Gel

Makes: single use
Shelf life: use immediately

1 teaspoon xanthan gum
2 teaspoons glycerine
2 teaspoons washing-up liquid
300ml/10½fl oz just-boiled water
1 teaspoon salt
5 tablespoons soda crystals

Is cleaning the oven your least favourite job? Mine, too, my friend, mine, too. I shan't admit how often I clean my oven (you might judge me), but let's just say it's not as often as I should. For me, the big problem is that I find the caustic fumes from conventional oven cleaners overwhelming. Instead I've come up with this oven-cleaning gel that does the trick without the noxious atmosphere. The gel ensures the cleaner clings to the sides of your oven overnight, allowing the soda crystals to really get in and help dissolve the cooked-on grease and grime. The result is a clean oven without the chemical stench.

Put the xanthan gum and glycerine in a large bowl and stir well to fully combine. Add the washing-up liquid and stir again.

Put the just-boiled water in a jug and add the salt and soda crystals. Stir until the crystals dissolve.

Pour the warm solution into the bowl with the xanthan gum mixture and use a hand-held blender to pulse for 1 minute, until fully combined.

To use

Switch off your oven at the socket and remove the racks from the inside. Wearing rubber gloves, use a sponge or scrubbing brush to apply the gel liberally to the surfaces of your oven, including the door.

Leave the gel on overnight, then in the morning, again wearing rubber gloves, use a scrubbing brush to give your oven a thorough clean. If burnt-on spots remain, sprinkle over some bicarbonate of soda to give you extra scouring power.

When you're satisfied, wipe the oven down with a clean, damp cloth, rinsing the cloth in fresh water as necessary.

⚠ Do not use on aluminium surfaces.

* Use this solution on the oven racks and trays, too, if you like.

Lemon Grove Microwave Cleaner

Burnt-on food in your microwave can seem a bit daunting to tackle, but this cleaner works its magic while you carry on with other tasks (or, my favourite, while you sit down and have a cup of tea). Want to know the very best part? It makes your kitchen smell lemony fresh, like you're sitting in a Mediterranean lemon grove!

Makes: single use
Shelf life: use immediately

300ml/10½fl oz just-boiled water
1 lemon, halved

Place the hot water in a microwave-safe bowl. Squeeze the juice from both lemon halves into the bowl of water, then pop in the lemon skins, too.

To use

Place the bowl in the microwave, and microwave on full power for 3 to 4 minutes. When the time is up, don't open the microwave door. Instead, leave it closed for up to 30 minutes to allow the lemony water to steam the grime from the inside of the microwave.

After this time, remove the bowl and wipe your microwave with a clean, damp cloth. Any debris should easily wipe off.

Burnt-pan Rescue Remedies

Food cooking doesn't come naturally to me. We joke in our house that on the nights when I cook everyone knows when dinner is ready as the smoke detector goes off. Needless to say, these burnt-pan cleaning recipes are ones that I find myself preparing often. They work a treat, but hopefully, you won't need to use them as much as I do!

Makes: single use
Shelf-life: use immediately

Stainless-steel pans:
Bicarbonate of soda (the amount will depend upon the size of your pan) or hydrogen peroxide 3% or 6%

Cast-iron pans:
1 tablespoon coarse salt, plus extra if needed
1 tablespoon olive oil, plus extra if needed

Stainless-steel pans
Sprinkle bicarbonate of soda over the bottom of your pan to cover, then add water to a 2.5-cm/1-inch depth. Place the pan on a high heat and bring to the boil. Reduce the heat to low, simmer for about 15 minutes, then use a wooden spoon or spatula to dislodge the burnt-on food. Remove the pan from the heat, discard the liquid and wash out the pan as normal.

Alternatively, pour enough hydrogen peroxide into the pan to completely cover the bottom. Leave to soak overnight, discard the liquid and wash as normal.

Cast-iron pans
Sprinkle the salt in the pan and add the oil. Using a damp cloth, gently scrub the pan in circular motions until clean. Add more salt and oil, if required, and keep scrubbing until the pan comes clean.

Nourishing Wood Butter

Wooden chopping boards and utensils need a bit of love now and again. All that cutting and stirring, then washing and wiping really takes it out of them. This wood butter moisturises and protects your wood and is really straightforward to put together using just two simple ingredients.

Makes: 125g/4½oz
Shelf life: See the use-by date
    on the coconut oil

25g/1oz beeswax pellets
100g/3½oz coconut oil

1 tin can
125g/4½oz lidded jar

Put the beeswax pellets in the tin can. Fill a small pan one-third full with water, then put the tin in the water, and carefully place on the hob on a low heat. Allow the water to heat up gently, but ensure none of it gets into the can, until the beeswax has completely melted. Add the coconut oil, and allow that to melt, too. Stir to combine, then pour into the jar, pop on the lid and leave the mixture to cool and solidify.

To use

Scoop out a dollop of the wood butter onto a clean cloth and rub the butter into your chopping board or wooden utensils. Leave the butter on overnight, then in the morning give your boards and utensils a good buff with a clean, dry cloth.

Zingy Chopping-board Deodoriser

Makes: single use
Shelf life: use immediately

1 tablespoon salt
½ lemon

I normally spray and wipe my wooden chopping boards, either with kitchen spray or hydrogen peroxide, but that doesn't always cut it when you've chopped something particularly odorous! (I'm looking at you, onion, garlic and fish.) Some stenches just do not leave the chopping board in a hurry, and the last thing you want is for them to transfer to the next meal you're preparing. So, for the really smelly stuff, I like to follow the zingy lemon deodorising method.

Sprinkle the salt on your chopping board and vigorously rub with the lemon half, gently squeezing as you go to release the juice.

Let the salt and lemon juice sit for around 10 minutes, before cleaning the board with a clean, damp cloth. To disinfect, simply spray some disinfectant (see pages 36–7) lightly over the board.

Kitchen Appliance TLC

With two kids my washing machine and dishwasher are barely turned off. These appliances are so much part of everyday life, it's easy to take them for granted. However, they do need some fairly frequent TLC, which means regularly giving them a good clean.

Of course, it seems absurd to clean the very machines we use for cleaning, but it does help to prolong their life, and ensures they always work at maximum efficiency.

Readily available cleaning solutions for washing machines and dishwashers (and kettles) often contain questionable ingredients and can be quite pricey, so I'm offering more natural solutions that use ingredients you may already have to hand.

Washing Machine

Washing machines need regular cleaning otherwise soap and limescale can build up causing all sorts of nasty problems, including odours on your clothes, and bacteria growth in the machine.

Thankfully, you won't need much elbow grease. Once every eight weeks add 500g/1lb 2oz of soda crystals to the empty drum and then, without adding any laundry to your machine, run the longest and hottest wash setting you can. This cleans out the drum and pipework good and proper.

You may need to repeat this step one more time if it has been a long time since you last cleaned your machine, and particularly if you've been using liquid laundry detergent and/or conventional fabric conditioners on low-temperature washes.

Limescale-busting
While soda crystals do a great job at removing gunk from your machine, they don't remove limescale. For that, run the same empty, long, hot wash, but this time with 500ml/19fl oz vinegar in the drum.

Unfortunately, you can't take a shortcut and run the soda crystals and vinegar at the same time, because the vinegar would just react with the soda crystals to form water and salt, which would do nothing to clean your machine. Pesky chemistry.

Between deep cleans
To help keep my machine clean in between deep cleans, every week I like to wash my towels at 60°C/140°F using my regular detergent. Washing at low temperatures all the time, while great for the environment, isn't too great for your machine, and can lead to musty smells. Washing towels on a hot wash once a week a) cleans your machine and b) cleans your towels – so it's all good!

Dishwasher

I advise giving your dishwasher a good old clean once every four weeks or so.

First, remove the filter and give it a scrub in hot, soapy water. Pop the filter back in, then fill two dishwasher-safe bowls with about 250ml/9fl oz white vinegar. Place one bowl in the middle of the top rack of the machine, and one in the middle of the bottom rack, making sure there's nothing else in the dishwasher. Run the dishwasher on the longest and hottest wash you can. That is it! You're done, without even breaking a sweat!

Kettle

If you live in a hard-water area, you'll be well aware of the need to regularly descale your kettle. Some people like using vinegar (equal parts white vinegar to water), but I've found you then have to boil your kettle a good few times with clean water afterwards to remove the vinegar smell. That's a bit too much effort for my liking, so here's my simple alternative.

Every month or so, simply add 1 tablespoon of citric acid to half a kettle of water, and switch on the kettle allowing it to boil the water and switch off (or, if it's a stovetop kettle, bring it to the boil and remove it from the heat). Don't empty the kettle straightaway, but leave it to sit for about 30–60 minutes, then pour away the contents and rinse well.

LAUNDRY
#dirtywashing

The laundry aisle in the supermarket can be an assault upon the senses. From the overwhelming choice of powders and liquids all claiming to be more effective than their competitors, to the overpowering aromas, and the new products we're told we must have – it's certainly not an aisle I like to hang around in.

Instead I make my own laundry products using simple yet effective ingredients, such as vinegar, lemon, and bicarbonate of soda. These aren't just better for humans and the environment, you'll soon see they can be much better for our clothes, too, helping to prolong their longevity.

Making your own laundry products is far cheaper than buying shop versions, too. From laundry detergent, to fabric conditioner, to fabric refresher sprays, I'm confident I've covered all the laundry eventualities you'll ever need!

Don't miss my handy list of natural stain-removal techniques to help tackle a wide range of stains. It's a well-thumbed page in our house...

LAUNDRY

Orange & Grapefruit Laundry Powder

Makes: 1.5kg/3lb 5oz
Shelf life: up to 12 months

500g/1lb 2oz soda crystals
500g/1lb 2oz borax substitute
250g/9oz bicarbonate of soda
2 bars (or 300g/10½oz) castile
   soap or vegetable-based soap,
   grated (see tip, below)
20 drops sweet orange essential oil
20 drops grapefruit essential oil

1.5kg/3lb 5oz lidded glass jar

Swapping from a commercial laundry detergent to a homemade one was the one change I was most apprehensive about. Decades of persuasive advertising from detergent companies made me feel my clothes wouldn't be clean without whatever special ingredients their detergents possess.

It turns out my fears were completely unfounded. This laundry powder works great at cleaning my clothes. My biggest critic is my one-year-old daughter. Still in cloth nappies, she has never shown signs of skin irritation as a result of the laundry powder, and I've noticed her nappies smell much fresher than they did before my switch. If that isn't success, then I don't know what is!

Mix together the soda crystals, borax substitute and bicarbonate of soda in a large bowl.

Put the grated soap in the bowl of a food processor and add a large scoop of the soda, borax and bicarbonate mixture. Add the essential oils, then blitz until you have a very fine powder.

Tip the blitzed powder into the bowl with the remaining soda, borax and bicarbonate mixture, stir to combine, and transfer to the airtight jar.

To use
Depending upon how soiled your clothes are, put 1–2 tablespoons of the mixture into the detergent tray of your machine per full load of laundry.

If you use a cold wash (lower than 30°C/85°F), soda crystals won't always fully dissolve. In this case simply dissolve the powder in a little hot water, then add that to your soap-dispensing drawer.

---

* To grate the soap, I find it easiest first to cut my soap into cubes using a hot knife, and then to use the grating disk on my food processor, but you can use a box grater if you prefer.

Lemon-scented Fabric Conditioner

Makes: 500ml/17fl oz
Shelf life: 5+ years

500ml/17fl oz white vinegar
30 drops lemon essential oil

500ml/17fl oz glass bottle or
    lidded glass jar

Fabric conditioner has a dirty secret: not all brands are vegetarian- or vegan-friendly. That's right: one ingredient found in certain brands is dihydrogenated tallow dimethyl ammonium chloride. In simpler terms: animal fat extracted from suet, the fatty tissues around the kidneys of cattle and sheep.

Meanwhile a study by the University of Washington found that certain chemicals in scented fabric conditioner are likely carcinogens, and that they contain allergens that can contribute to eczema. And all that's on top of the fact that conventional fabric conditioner lessens your laundry's ability to properly absorb water and laundry powder (see page 60). This means that the more you use conventional fabric conditioner, the less well your clothes will respond to washing, increasing the likelihood that they will lock in bad odours.

I use the following recipe as an alternative. White vinegar makes for a good natural fabric conditioner because it cuts through soap residue (it's the excess soap in your laundry that makes your clothes and towels feel rough). Crucially, it won't interfere with the absorbency of your laundry. Remember, vinegar dries without an odour, so once your clothes are dry, you will have softened, clean-smelling clothes and a delicate and clean aroma on your laundry, without a hint of eau de vinegar!

Simply fill your bottle or jar with the vinegar, add the essential oil, stir to combine and screw on the lid tight .

To use
Shake the fabric conditioner, then for a full load of washing, simply fill up to the line on the fabric-conditioner drawer of your machine (for a half-load, use half the amount, and so on).

#ortrythis Lemon essential oil complements the washing powder recipe on page 60, but sweet orange, or your own personal favourite, would work just as well. Alternatively, you can skip the oil altogether, and just use vinegar for a scent-free conditioner.

It's a White Wash

The one downside to using homemade laundry powder is that whites need a little bit of help to stay white. Most conventional detergents use optical brighteners – synthetic chemicals that are, on the face of it, quite clever. They transform ultraviolet light waves to enhance blue light and minimise yellow light, which has the effect of making your clothes appear whiter. Notice my choice of words – appear whiter. They don't actually get your clothes any cleaner.

However clever they may be, though, they aren't good for humans or the environment, having been linked with affecting the human reproductive system and damaging aquatic life (because they aren't biodegradable). So, here are a few natural alternatives to optical brighteners to keep up your sleeve.

Sunlight and fresh air
Dry your clothes outside whenever you can. Sunlight is the very best way to keep clothes white, and it's free!

Lemon juice
Put 100ml/3½fl oz of lemon juice in the pre-wash compartment

Hydrogen peroxide
Put your usual 1–2 tablespoons of laundry powder into your washing machine, then add 250ml/9fl oz of 3% hydrogen peroxide in the pre-wash compartment.

Natural Stain Removers

I'd like to say the reason I have amassed many natural stain-remover techniques is because of my two young daughters, who get into the kind of messes that only young kids can. The truth is I'm a tea-swilling, homemade-soup-slurping, pottering-in-the-garden kind of lady, who is more than capable of getting into my own messes! Whatever your reason for stained clothes, I'm sure I've got a method here for you.

As with any stain remover, before trying out any of these tips on your clothes or upholstery, I recommend spot testing in an inconspicuous area just in case.

Baby poo

This is quite a tenacious stain, and for some reason most baby clothes are either white or pale-coloured, meaning stains are inevitable. My top tip is to mix 300g/10½oz of soda crystals into 570ml/20fl oz of hot water, and pre-soak the clothes for 1 hour before washing. If the stains persist, hang the clothes (or cloth nappies) outside on a sunny day as sunlight is particularly effective at bleaching out poo stains.

Biro

Enzymes in milk are great at shifting biro stains. Simply soak the item of clothing in a little milk for about 3 hours, then wash as normal. If you have biro-stained upholstery, try soaking a cloth in milk and use it to rub off the stain.

Blood

Either pre-soak your item of clothing in heavily salted cold water and wash as normal, or soak it in a mixture of 1 litre/35fl oz of hot water mixed with 4 tablespoons of soda crystals, then wash as normal. Either option, depending on what you have to hand, should remove the most stubborn of blood stains.

Candle wax

Got some candle wax on your best table cloth? Try placing brown paper on top of the the waxy stain and then ironing the paper with a warm iron. This should draw the melted wax out of the fabric. You may have to repeat this method a few times for particularly stubborn stains.

Chewing gum

Pop your piece of clothing in the freezer and leave it for a few hours. The gum should then be quite hard and brittle, so that you can scrape it off with a knife.

Cooking fat

Create a paste of equal parts bicarbonate of soda or soda crystals and water. Spread this over the stain, then leave on the paste for 30 minutes and wash as normal.

Crayon

Put the item of clothing into the freezer to harden the crayon, then scrape off with a knife. Then, place the stained area of clothing between two clean paper towels, and press with a warm iron. This should transfer the residual crayon wax to the paper towels. You may need to do this a few times. Finally, wash as normal, adding a large spoonful of soda crystals in the drum to help remove any residue.

Lipstick

Remove the crusts from a slice of white bread and roll the doughy part into a ball (trust me on this!). Use the ball of bread to blot the lipstick stain, which should lift from the clothes, then wash as normal.

Mud

Pre-soak muddy clothes in a bucket of warm water with 4 tablespoons of bicarbonate of soda or soda crystals for 3 hours. Then, wash as normal. Don't be tempted to leave the clothes sitting muddily before you clean them. My technique works best if you soak them straightaway.

Perspiration marks

Soak your clothing in white vinegar for at least 30 minutes to 1 hour, rinse in water and then wash as normal.

Red wine

Make a paste of equal parts bicarbonate of soda and water and apply to the stain. Leave for 2–3 hours, rinse under a tap, then wash as normal. Sparkling or soda water may also help remove a red-wine stain if bicarbonate of soda doesn't cut the mustard.

Rust

Pour 1 tablespoon of either vinegar or lemon juice onto the stain, then blot with a clean, white cloth. If you can, hang out the item of clothing in the sun for a few hours, then wash as normal.

Tea & coffee

Mix 300g/10½oz of soda crystals into 570ml/20fl oz of hot water. Soak your clothes in the mixture for 1 hour, then wash as normal.

Disappearing Stain Spray

Makes: 100ml/3½fl oz
Shelf life: about 4 weeks

2 tablespoons washing-up liquid
100ml/3½fl oz 3% hydrogen
　peroxide
5 drops lemon essential oil
5 drops eucalyptus essential oil

100ml/3½fl oz amber-coloured
　glass spray bottle

As a working mum sometimes I don't have time to mess around with soaking and scrubbing. If you're in a bit of a rush and just want to be able to spray and go, I recommend this recipe. I make up a small quantity at a time, as hydrogen peroxide eventually loses its potency when mixed with other ingredients. I've found this combination works well on whites and coloured fabrics, but do spot test before use. Treating the stain as soon as it occurs is the most effective method. Older 'set' stains may require repeated application.

Put all the ingredients in the spray bottle, pop on the pump lid and shake well to combine.

To use

Shake well and spray liberally onto your stain. Rubbing in the spray with an old toothbrush or scrubbing brush helps. Leave for at least 15 minutes, rinse off, then wash as normal.

---

* Drying outside in the sunshine will help to enhance the stain-removal effects.

Bergamot & Grapefruit Fabric Refresher

Makes: 500ml/17fl oz
Shelf life: about 8 weeks.

250ml/9fl oz witch hazel or vodka
250ml/9fl oz cooled boiled water
¼ teaspoon glycerine
40 drops grapefruit essential oil
40 drops bergamot essential oil

500ml/17fl oz glass spray bottle

If you've ever bought a piece of vintage clothing, you may well be familiar with the dreaded 'vintage' smell. I've come up with a delicious-smelling bergamot and grapefruit fabric refresher spray that helps banish stale smells with ease, and without any of the dodgy stuff that lurks in conventional fabric freshener.

The real secret is the witch hazel or vodka, which helps to remove the stench as they evaporate from the fabric. If you use vodka, I promise your clothes won't smell like a pub! Vodka dries without even a hint of an odour, but if you're worried, witch hazel is equally as effective. Always patch test on an inconspicuous area first.

Put all the ingredients in a spray bottle, pop on the pump lid and shake well to combine.

To use
Shake, then spray liberally onto the fabric and leave to dry. For extra-stinky items, saturate with the spray and hang outside to dry.

BATHROOM
#makeitsparkle

There are very few people I've come across who find cleaning the bathroom the most joyous task in the world. However, you can bring a little bit of joy to the proceedings by making your own bathroom cleaning products. If fizzing toilet cleaners don't bring a smile to your face, then surely a vodka-based daily shower spray will?

At the very least you can take satisfaction from the simple fact that homemade cleaning products are kinder to the environment and yet still leave your bathroom squeaky clean.

Unlike bleach, these products won't harm the friendly bacteria in your septic tank, if you have one. Nor will your bathroom smell like a swimming pool. In fact some of the products may actually smell good enough to eat (particularly the grapefruit and ginger foaming hand soap), but I wouldn't recommend it!

BATHROOM

Tea Tree & Peppermint All-purpose Bathroom Spray

If you're pushed for time, feel free to use the vinegar-based kitchen cleaning spray to clean your bathroom, too. However, if you're vinegar adverse, or looking for a specific bathroom cleaner, do try this. It's my favourite for scrubbing down the sink, toilet, bath and tiles, and anywhere that needs a good once over. It has excellent cleaning, deodorising and antibacterial properties.

Makes: 500ml/17fl oz
Shelf life: 4–8 weeks

500ml/17fl oz cooled boiled water
1 teaspoon bicarbonate of soda
1 teaspoon borax substitute
2 teaspoons liquid castile soap
¼ teaspoon glycerine
15 drops tea tree essential oil
30 drops peppermint essential oil

500ml/17fl oz glass spray bottle

Pour the water into a jug, then add the bicarbonate of soda and borax substitute. Stir until fully dissolved (dissolving completely will prevent it clogging up the bottle's spray mechanism). Add the remaining ingredients, mix well, and transfer to your spray bottle.

To use
Spray and then wipe away with a clean cloth, just as you would with any bathroom cleaner.

#ortrythis For an added cleaning boost in particularly grubby bathrooms, swap the bicarbonate of soda for an added teaspoon of borax substitute.

Lavender Shower Spray-n-Go

Makes: 500ml/17fl oz
Shelf-life: about 8 weeks

350ml/12fl oz just-boiled water
1 tablespoon borax substitute
150ml/5fl oz vodka
10 drops tea tree essential oil
25 drops lavender essential oil

500ml/17fl oz glass spray bottle

I used to make a shower spray with vinegar and essential oils, which was great at dealing with daily soap residue. The problem with it was that the smell was overwhelming in my tiny, windowless bathroom. Vinegar dries odourlessly, which is fine when you can spray and wipe down. With a daily shower spray, however, you just want to be able to spray and go. So, I've come up with a vinegar-free version that will help keep your shower cubicle fresh between thorough cleans.

Pour the just-boiled water into a jug, then add the borax substitute and mix well until fully dissolved. Stir in the vodka and essential oils and transfer the mixture into a spray bottle. Pop on the pump lid and shake well to combine.

To use
Shake well, then spray liberally all over the shower cubicle after every shower. There's no need to rinse after spraying, just leave to dry.

Supersoft Citrus Shower & Bath Scrub

For acrylic baths and shower trays, this gentle, soft scrub is great at cleaning up soap scum, tidemarks, dried-on toothpaste and anything else your bathroom can throw at you. I'd go as far to say it's one of my favourite homemade cleaning products. See for yourself!

Makes: 250g/9oz
Shelf life: about 8 weeks

200g/7oz bicarbonate of soda
1½ tablespoons liquid castile soap
1 tablespoon cooled boiled water
5 drops grapefruit essential oil
5 drops lemon essential oil
5 drops sweet orange essential oil

250g/9oz lidded glass jar

Put the bicarbonate of soda in the jar, then add the liquid castile soap and stir well. Next, add the water and oils and stir to form a smooth paste. It shouldn't be runny – you should be able to form a clump in your hand. If the mixture is too dry, add a tiny bit of water at a time until you have the right consistency. If it's too wet, add a bit more bicarbonate of soda. When you're happy, place the lid on the jar and secure tight.

To use
I use this scrub to give my sink, shower and bath a good, old scrub. My tool of choice is a small wooden scrubbing brush, intended for pots and pans. I wet the brush, then spoon a small amount onto the bristles and scrub the surfaces in circular movements. Don't forget to rinse down once everything is sparkling – this is especially important in the bath to avoid that gritty feeling when you get in for a soak!

Shower screens may need a final wipe down with a squirt of window cleaner (see page 100) to avoid smears, if you're worried about that sort of thing.

Miracle Tile & Grout Paste

Makes: 230g/8½oz
Shelf life: about 8 weeks

100g/3½oz bicarbonate of soda
50g/2oz borax substitute
2 tablespoons liquid castile soap
10 drops lemon essential oil
5 drops tea tree essential oil
4 tablespoons cooled boiled water

250g/9oz lidded glass jar

Grout. Such an unpleasant-sounding word, and cleaning the stuff is such an unpleasant job, don't you think? Until now, that is. This cleaning paste works while you put your feet up!

The formula is bleach-free, which means it is especially good for cleaning coloured grout (bleach can cause discolouration). The grout in our bathroom is grey – and this paste makes relatively light work of the horrible orange grime that seems to thrive whenever I've been a bit lax on the tile-cleaning front (well, it happens to the best of us). Truly a miracle!

Put all the ingredients except the water into the jar and use the handle of a wooden spoon to mix well.

Add the water, a little at a time, until you have a paste the consistency of cupcake frosting. If you need to add more water that's fine; if you feel like you've added too much water, simply add a little more bicarbonate of soda.

To use
You'll need to wear rubber gloves. Wet a pan brush or old toothbrush and coat the bristles in some of the paste. Scrub the tiles and the grouting in-between. If you're cleaning wall tiles, this bit is quite messy, but don't worry – the results are worthwhile.

Leave the formula on your tiles for about 15 minutes or so. (I like to use this time to sit back and have a cup of tea, and later I'll pretend to everyone else that I was scrubbing the bathroom the whole time, but it's your call.) Whatever you do, don't get carried away and leave the paste on for longer than 30 minutes, as when borax substitute hardens it's really difficult to get off!

After the 15 mins or so, wet the bristles of an old toothbrush and scrub the grout thoroughly, then rinse or wipe clean with fresh water.

Borax substitute has a habit of clumping together, but don't worry if your mixture is lumpy – the paste will do the job even if it isn't perfectly smooth.

Eucalyptus, Grapefruit & Ginger Air Freshener

Bathrooms sometimes need a little helping hand to smell fresh. This air freshener is deliciously scented to make your bathroom smell like a fancy spa without the overpowering aroma.

Makes: 500ml/17fl oz
Shelf life: about 8 weeks

250ml/9fl oz witch hazel
250ml/9fl oz cooled boiled water
¼ teaspoon glycerine
40 drops grapefruit essential oil
40 drops ginger essential oil
40 drops eucalyptus essential oil

500ml/17fl oz glass spray bottle

Put all the ingredients in the spray bottle, screw on the pump lid and shake well to combine.

To use
Shake well, then spray liberally (at least five pumps) into the room.

Lemon & Lavender Clinging Toilet-bowl Cleaner

A lot of people seem to prefer homemade cleaning products if they behave in exactly the same manner as the shop-bought cleaners we are used to. With toilet cleaners that means a formula that clings to the bowl as you squirt. When I came up with this clinging toilet-bowl cleaner I decided I'd hit upon the holy grail of natural cleaning. The magic is in the vinegar, which cuts through limescale, and in the oils and soap which help clean and freshen as the solution clings to the bowl and rim. High fives all round!

Makes: 500ml/17fl oz
Shelf life: about 8 weeks

1 teaspoon xanthan gum
2 teaspoons glycerine
2 teaspoons washing-up liquid
200ml/7fl oz white vinegar
35 drops lemon essential oil
25 drops lavender essential oil
15 drops tea tree essential oil
300ml/10½fl oz cooled boiled
    water

500ml/17fl oz squeezy plastic
    bottle

Put the xanthan gum and glycerine in a measuring jug and stir for a few minutes until the xanthan gum is fully dissolved. (This step is vital – if you skip it, the gum won't mix well and you'll end up with a gloopy mess.)

Add the washing-up liquid to the mixture, and stir well to combine. Then add the vinegar and essential oils along with the water. Use a hand blender to pulse the mixture for 1 minute, until fully combined. Decant the toilet cleaner into your squeezy bottle.

To use
Shake well, then squirt the mixture around the toilet bowl, just as you would conventional cleaner, and leave it to sit in the bowl for at least 15 minutes before giving your toilet a good scrub.

* It's more usual to use a glass bottle for any mixture that includes essential oils, but in this case I prefer plastic. A rinsed-out washing-up liquid bottle is perfect. Just check your bottle on a regular basis for any sign of degradation as a result of the essential oils.

Lemon & Geranium Toilet-bowl Scrub

If your toilet bowl is in need of a good scrub, this will sort it right out with its stain-busting powers of citric acid and borax substitute. It's especially great on limescale. And while some conventional toilet cleaners smell like they might just burn the hairs right out of your nostrils, this one will scent your bathroom with the delicate aroma of lemon and geranium.

Makes: 450g/1lb
Shelf life: about 8 weeks

150g/5½oz bicarbonate of soda
150g/5½oz borax substitute
150g/5½oz citric acid
25 drops lemon essential oil
15 drops geranium essential oil

450g/1lb lidded glass jar

In a bowl combine the bicarbonate of soda, borax substitute and citric acid (take particular care when adding the citric acid, as the concentrated nature of the powder can irritate the skin, eyes and airways if spilled). Add the essential oils and stir well to combine, then carefully decant into the jar.

To use
Take a generous spoonful of scrub and sprinkle it around the toilet bowl. It will start to fizz gently. Give it a few minutes, then scrub the bowl and rim with a toilet brush, and flush to rinse.

* Make sure your jar is airtight and keep the scrub away from moisture to maximise the shelf-life.

Bergamot & Rosemary Toilet-fresh Tablets

Makes: about 6 tablets
Shelf life: up to 4 weeks

500g/17oz bicarbonate of soda
125g/4½oz citric acid
20 drops grapefruit essential oil
20 drops bergamot essential oil
20 drops rosemary essential oil
2 tablespoons witch hazel

1 small spray bottle
1 silicone 6-hole muffin tray
1 lidded glass jar

There aren't many five-year-olds who beg to clean the toilet, but mine does! She drops a tablet into the bowl and loves watching it fizz up as it sets to work cleaning all the nooks and crannies. The magic is in the citric acid, which helps to dissolve the limescale that can build up in your toilet, while the bicarbonate of soda acts as a mild abrasive and neutralises bad odours.

Put the the bicarbonate of soda and the citric acid in a bowl and mix, then add the essential oils, and mix again.

Pour the witch hazel into a small spray bottle, and lightly mist the mixture of dry ingredients with one or two sprays of witch hazel at a time. The trick is to moisten the mixture enough to be able to mould it, but not so that it is so wet it reacts in the bowl. Every time you spray, mix well and test to see if the mixture will form a clump in your hand. It shouldn't take many sprays, about 6 or 7, to get the mixture to the right consistency – so don't get carried away.

Once the mixture clumps together, use a spoon to pack it into your silicone mould. I use a silicone muffin tray with 6 holes in it, to make 6 large tablets. Make sure you press firmly down with the back of your spoon, so that the mixture is tightly packed.

Leave the tablets to dry overnight, then the next day pop them out and store them in an airtight jar.

To use
Drop 1 tablet into your toilet bowl. Leave it to sit for at least 30 minutes, or ideally overnight (for the best results), then scrub with a toilet brush and flush to rinse. For stubbornly stained toilets, use two tablets; and for really stubbornly stained toilets, repeat every day for a week.

* Don't use these tablets in your cistern, only in the toilet bowl, and wait until the tablets have stopped fizzing before you flush for maximum cleaning efficiency.

Tangy Grapefruit & Ginger Foaming Hand Soap

Makes: 300ml/10½fl oz
Shelf life: up to 8 weeks

100ml/3½fl oz unscented liquid
    castile soap
5 drops grapefruit essential oil
5 drops ginger essential oil
200ml/7fl oz cooled boiled water

300ml/10½fl oz foaming hand
    soap dispenser

Think you don't have time to make your own cleaning products? Think again: you'll be able to put this foaming hand wash together in less time than it takes to make a cup of tea!

It's a great idea to make your own hand soap. Conventional versions often contain triclosan. This is an antibacterial agent, which has been banned in the USA, as it may be harmful to human health. Antibacterial agents, such as triclosan, can also promote antibiotic resistance, giving you another good reason to ditch the shop-bought soaps and make your own.

If you're worried about germs, a study conducted at Korea University in 2015 found that antibacterial handwash is no better than regular soap at killing germs and viruses from hands during washing.

Use an empty foaming hand-wash container for this recipe. One bottle of liquid castile soap will make several refills and using the unscented version allows you to personalise the fragrance using your favourite essential oils. The combination given here is one of my favourites – grapefruit and ginger, which gives a really zingy and uplifting scent. Of course, you can use pre-scented liquid castile soap if you prefer.

Fill the dispenser with the liquid castile soap and essential oils.

Add the cooled boiled water, and screw on the dispenser. Shake well to combine.

To use
Dispense 1 or 2 pumps on to your hands, wash as normal and rinse for a tangy clean!

#ortrythis To turn this into a moisturising hand wash, add 50ml/2fl oz of vitamin-E oil to the liquid castile soap and essential oils before adding 150ml/5¼fl oz water.

Forest Fresh Solid-Floor Cleaner

I used to use bleach to clean my solid floors (such as linoleum, vinyl and tiles). It did a fine-enough job, but the smell! My house ended up stinking like a swimming pool, and made me feel nauseous. The result was that I avoided cleaning the floors as much as possible. These days I've opted for a natural solution that leaves my home smelling like a forest instead of a municipal pool, which is always a bonus.

Makes: about 4 litres/7 pints
    (single use)
Shelf life: use immediately

150ml/5fl oz witch hazel
1 tablespoon borax substitute
40 drops pine essential oil
3–4 litres/5–7 pints very warm
    water

Add the witch hazel, borax substitute and essential oil to the warm water and stir well, ensuring the borax substitute dissolves fully.

To use
Dip your mop into the mixture. Wring out as much liquid as possible, then wipe over your floor as normal.

GENERAL
HOUSEHOLD
#allthenooksandcrannies

Now that we've tackled the kitchen, laundry and bathroom you might be wondering what's left to possibly clean? Well, for this chapter I've compiled my favourite general household cleaning recipes – from floors, to carpets, to woodwork, to windows – to really make your home sparkle.

I don't think I've ever had a home that doesn't have a few awkward nooks and crannies that get forgotten about. Who doesn't have a dark corner where mildew might grow? This chapter has a formula for those hidden horrors, and even a freshener for a nappy bin.

Finally, if you're looking to freshen up your air a little, I've got a few natural deodorising techniques up my sleeve for you. There's the super-quick, such as a scented Lemon, Grapefruit & Geranium Deodoriser, to slightly more involved beeswax candles, perfect for setting a mood when you have guests, or even to give as gifts.

GENERAL HOUSEHOLD

Shaken-not-Stirred Mirror & Window Cleaner

A lot of people recommend straight vinegar for cleaning windows, but I have to disagree. I've tried out a whole host of different methods, including just straight vinegar, and this recipe, which combines vinegar with other ingredients, is by far is the best.

It's secret? Vodka, which if you're reading the book from start to finish, by now won't surprise you! The other magic ingredient, though, is cornflour. Yes, the stuff you use for thickening gravy. The vinegar cuts through dirt and grease, the vodka helps eliminate streaks and remove any remaining residue, and the cornflour acts as a very gentle abrasive to help break down any stubborn dirt and grime with very little elbow grease. Try this formula and I swear you won't look back.

Makes: 500ml/17fl oz
Shelf life: up to 8 weeks

500ml/17fl oz just-boiled water
    allowed to cool slightly
1 tablespoon cornflour
2 tablespoons vodka
2 tablespoons vinegar
10 drops lemon essential oil

500ml/17fl oz glass spray bottle

Put the water and cornflour in a measuring jug and stir well until fully dissolved. Add the remaining ingredients, then mix well and decant into the spray bottle and secure the lid.

To use

Shake well, then spray on to glass surfaces or mirrors and rub with a clean, dry cloth in circular motions. Once you have thoroughly cleaned the whole surface, take another clean, dry cloth – or a crumpled-up piece of newspaper – and buff dry. At first the surface will look misted, but once enough of the liquid has dried up, you'll be left with spotless and gleaming glass.

⚠ Do not use this glass cleaner to clean car windows, as if you get the mixture anywhere else, the vinegar may damage the car's paintwork, especially if your car has metallic paint.

Citrus & Pine Furniture Polish

Makes: 300ml/10½fl oz
Shelf life: 5+ years

100ml/3½fl oz jojoba oil
200ml/7fl oz white vinegar
¼ teaspoon glycerine
30 drops lemon essential oil
10 drops sweet orange
  essential oil
10 drops pine essential oil

300ml/10½fl oz glass spray bottle

A can of conventional furniture polish has a long list of ingredients, including butane, paraffin, and other waxes derived from fossil fuels, and sometimes formaldehyde – yes, the carcinogenic stuff embalmers use.

All you need to make your own natural furniture polish is vinegar to clean, jojoba to nourish and protect, and a few essential oils to add some scent. I think a fresh-smelling citrus and pine blend suits this polish really well.

Technically, jojoba isn't an oil but a liquid wax, which leaves a lovely finish on wooden surfaces. It has a really long shelf life, so that your bottle of furniture polish can last and last. This is perfect if dusting is something you do only when you have visitors coming over. Or is that just me?

Put all the ingredients in the spray bottle, pop on the pump lid and shake well to combine.

To use
Shake well, then spray the polish onto a clean, dry cloth, then gently buff in circular motions to bring up the shine in your wood.

#ortrythis If you're in a pinch and don't have any jojoba oil to hand, feel free to substitute it with olive oil. This has a tendency to go rancid fairly quickly, so make a single quantity – mix 1 tablespoon of olive oil with 2 tablespoons of white vinegar and 4 drops of essential oil. You can skip the glycerine in this case.

Pine-scented Wood & Laminate Floor Spray

Makes: 500ml/17fl oz
Shelf life: about 8 weeks

3 tablespoons white vinegar
500ml/17fl oz cooled boiled water
40 drops pine essential oil
2 teaspoons washing-up liquid

500ml/17fl oz glass spray bottle

Wood and laminate floors need a bit more cleaning care than floors that are tiled or covered with vinyl.

Soaking them in water can, over time, warp the wood or laminate and cause discolouration. So, rather than busting out the bucket and mop, I've put together this floor cleaner that you spray directly onto the floor, then wipe over with a damp mop or cloth. This gently cleans and leaves your floors with a lovely streak-free shine.

Combine the vinegar, water and oil in the spray bottle, then add the washing-up liquid and stir very gently using the long handle of a spoon – you'll need to be gentle as you don't want the washing-up liquid to foam up. Screw on the pump lid and then swill gently to combine.

To use
Spray the cleaning solution to form a light film on your floor, then rub with a very slightly damp mop or cloth. Continue to spray and wipe, until your whole floor is clean.

Lemon, Grapefruit & Geranium Deodoriser

Makes: 250g/9oz
Shelf life: about 3 months (single use)

250g/9oz bicarbonate of soda
About 15 drops each of lemon, grapefruit and geranium essential oils

250ml/9fl oz Kilner jar with a two-piece lid
Small scrap of breathable fabric, such as cotton or linen, cut so that it is slightly larger than your jar opening

What are your least favourite household products? Mine are plug-in air fresheners. I find the smell completely overpowering. And not just that – reports from Public Health England worryingly warn that plug-in air fresheners produce considerable levels of formaldehyde, a carcinogen closely linked with cancers of the nose and throat.

Thankfully, if you want to freshen up a stale part of your house, I've got a quick, easy and thrifty recipe, free from any questionable ingredients.

Put the bicarbonate of soda in your Kilner jar, add the essential oils and stir gently to combine. (You can add fewer or more drops of essential oil, if you prefer.)

Place the fabric over the jar, and secure it using the metal screw band from the Kilner lid. You don't need the vacuum seal, so keep it safe somewhere in case you ever need to use the jar for a different purpose.

To use
Place your jar in the area you want to freshen, but out of the reach of pets and children. The room deodoriser will freshen a room or cupboard for up to 3 months, but if the scent starts to wane before then, simply shake or stir the contents to refresh.

#ortrythis Here are some wonderful alternative essential oil blends to try:
• 15 drops each bergamot, orange and ylang ylang essential oils
• 20 drops each lemon and vetiver essential oils
• 20 drops each basil and lemon essential oils

Citrus Upholstery Deodorising Powder

If you have pets you'll know that carpets, rugs and sofas (if your animals are allowed on them) tend to need a good freshen-up on a regular basis. Unlike some commercial alternatives, this super-quick formula won't irritate your furry friends' delicate skin, nor yours either. I love a good multi-tasker, and this powder is also great for freshening up a mattress. If your mattress smells a bit stale, just sprinkle over a smattering of this – it works a treat.

Makes: 250g/9oz
Shelf life: up to 3 months

250g/9oz bicarbonate of soda
20 drops lemon essential oil
20 drops grapefruit essential oil

250g/9oz lidded glass jar or flour shaker

Pour the bicarbonate of soda into the glass jar or flour shaker and add the lemon and grapefruit essential oils, then stir gently to combine.

To use

Sprinkle the powder liberally onto your upholstery, carpet, rug or mattress, and leave it for at least 30 minutes, then vacuum thoroughly, ensuring you vacuum up all the mixture.

#ortrythis You can also use this powder to deodorise stinky shoes – simply sprinkle inside the shoes, leave overnight and empty out the next morning!

Sweet & Simple Beeswax Candles

Makes: 4–5 candles
Shelf life: 5+ years

Beeswax pellets (the amount
    depends on your containers, see
    method)

4–5 small metal or ceramic
    containers
1 tin can
4–5 bamboo skewers
4–5 petroleum-free candle wicks

I love nothing more than burning scented candles to freshen the air in my house. The trouble is that conventional candles are made using petroleum-based products, such as paraffin wax. When it burns, paraffin wax releases chemical nasties such as benzene and toluene – which isn't a particularly cleansing thought when you're trying to refresh the air. Thankfully, making your own non-toxic candles is simple.

Put some newspaper down to protect your work surface, and lay out the candle containers.

To gauge the amount of beeswax you need per container, fill one container with pellets. You will need double the amount once they are melted down, so, if your container holds 50g/2oz of pellets, you'll actually need 100g/3½oz for each container. (For 4 containers that means 400g/14oz, for 5 that means 500g/17oz.)

Pour the pellets into the tin can. Place the can into a small saucepan of boiling water and adjust the heat so that the water is at a rolling boil, but not so ferocious that any water gets into the can.

Stir continuously using one of the bamboo skewers to help the wax break down and liquefy. It should take about 15–20 minutes altogether.

Once the wax has melted, turn off the heat, and, using an oven glove, very carefully lift the hot can out of the water, then slowly pour equal amounts of the wax into each container. Take care – the wax will be extremely hot.

When each container is filled, add your wick – it may need supporting with a skewer until the wax starts to firm up. Simply lay the skewer across the top of the container and drape the wick over it.

Once the wax has hardened, trim the exposed section of your wick to no more than 1cm/½in in length.

⚠ Always place your candle on a heat-resistant surface when it's alight, and never leave it to burn unattended.

#ortrythis To make scented candles, add around 40 drops of your choice of essential oils to the pellets as you're melting them down. For a vegan candle, swap the beeswax pellets for soy wax flakes.

Sticky Stuff Remover

Makes: enough for 1 jam-jar label
   (single use)
Shelf life: use immediately (or
   by the use-by date on the
   coconut oil)

1 tablespoon virgin (not
   fractionated) coconut oil
1 tablespoon bicarbonate of soda

I've got a confession to make: I hoard glass jars. I love making jams and preserves, so I keep almost every jar that comes into my house. However, what I don't love is trying to remove the original sticky labels. Once upon a time I used to buy a special spray to do the job, but now I throw together a simple paste. It makes easy work of removing sticky labels and price tags, along with any other sticker you'd care to mention. Give it a go the next time you're cleaning jars for making jam; you'll thank me for it!

Place the coconut oil in a small bowl. If it needs liquefying, blast it on full power in a microwave for 5–10 seconds. Then, mix in the bicarbonate of soda until fully combined and leave the mixture to solidify a little before using.

To use
Apply a little of the paste to a cloth and rub it onto the surface of the label you want to remove. Leave for 1 hour, then peel away the label, and scrub with a fresh cloth to remove the remaining adhesive.

* You can make larger quantities of this paste, if you prefer. Just be sure to use it by the use-by date on your coconut oil.

Wood-burning Stove Cleaner

Makes: as needed
Shelf life: none

A small bowl of water
Ash from your fire

Got a wood-burning stove? Great! This means you pretty much have everything you need to keep the glass door of it sparkling clean. What am I talking about? I'm talking about ash.

It may seem completely counterintuitive to clean your wood-burning stove with the dirty ash from inside the stove itself, but ash is a mild abrasive and quite alkaline, making it a great cleaner. Historically, all the best housekeepers saved the ash from the fires to clean all round the home. They knew a thing or two! (The following technique works brilliantly on the glass doors of ovens, too.)

Scrunch up a piece of newspaper and dip the paper in the bowl of water. You want the paper to be moist, not drenched.

Next, dip the dampened newspaper into the ash from the ash box of your wood-burning stove (making sure the ashes aren't hot, of course), and use the dampened and dipped newspaper to scrub the glass until clean. Wipe over with a clean, damp rag, then scrunch a second bit of newspaper, and use this to buff the glass dry.

Peppermint Nappy Bin Discs

Makes: about 8 discs
Shelf life: up to 1 month in a bin
    with a liner; up to 3 months in
    a sealed jar

300g/10½oz bicarbonate of soda
40 drops peppermint essential oil

8-hole metal fairy cake tin
greaseproof paper
1 large lidded glass jar

If you use cloth nappies for your baby, then the nappy bin is never going to be the freshest place in your home. However, you can freshen it up a little, and take the edge off each time you lift the lid, with the help of good-old bicarbonate of soda.

Just one piece of advice: make sure you add the essential oils after you've baked the discs, as instructed. The first time I made these I mistakenly added the oils pre-baking, and let's just say I not only managed to freshen my nappy bin, but the whole house!

Preheat your oven to 180°C/350°F/gas mark 4.

Mix the bicarbonate of soda with 3 tablespoons of water to form a thick paste.

Line each hole in the metal fairy cake tin with greaseproof paper, and spoon your mixture evenly into each hollow. Place the tray in the oven and bake for about 20 minutes until dry and firm.

Remove the tray from the oven. Allow the discs to cool, then carefully remove each disc from the fairy cake tin. (You can leave the discs part-wrapped in their greaseproof paper.)

Add 5 drops of peppermint essential oil to each disc, then place the discs in the jar, seal and use as needed.

To use
I use a nappy bin that has a mesh liner, so I place 1 disc, paper and all, into the bin, then I put in the liner so that the disc and nappies don't come into contact with one another. Used in this way, the discs release their fragrance for about 1 month.

If you don't use a nappy bin liner, remove the disc from the paper wrapper, and place it in the bin. With the paper removed, when you tip the reusable nappies into the washing machine, you can tip in the disc, too. It will dissolve in the wash, helping to freshen the cloth nappies as they clean. You will have to add a new disc every time you empty the nappy bin.

Carpet & Upholstery Spot Cleaner

This spot cleaner helps remove almost any stain from carpet and upholstery. You might notice that this recipe is similar to the stain-remover spray for clothing (see page 68). The difference is that instead of the essential oils, I've used bicarbonate of soda. This is because you can't put your carpet in the washing machine, so you need something that will deodorise on-the-job. Do a patch test on an inconspicuous area first.

Makes: single use
Shelf life: use immediately

1 teaspoon bicarbonate of soda
1 teaspoon washing-up liquid
1 tablespoon 3% hydrogen
    peroxide

1 small glass jar or bowl

Put the ingredients in small bowl and mix well to combine.

To use
First, put on rubber gloves. Then, pour 1 teaspoon of the mixture onto your stain and scrub with an old toothbrush.

Add more of the stain-removing mixture as required, then when the stain has gone, blot the area with a wet cloth to remove any excess. Allow to dry, then vacuum to remove any bicarbonate of soda residue – an important step if you don't want to leave a white powdery residue on your carpet!

* Dark-coloured washing up liquid won't stain light-coloured carpet, but if you are in any way concerned, use a pale-coloured or clear washing up liquid.

Mould & Mildew Buster

Makes: 100ml/3½fl oz
Shelf life: up to 8 weeks

100ml/3½fl oz cooled boiled water
25 drops tea tree essential oil
10 drops lemon essential oil
  (optional)
10 drops rosemary essential oil
  (optional)

100ml/3½fl oz glass spray bottle

Here's something you might not know – mould can grow deep roots when it appears on porous surfaces, such as wood or breathable walls. But there's an easy and effective way to kill mould and mildew, right down to the roots, using tea tree essential oil. This oil is powerful stuff, and its small molecules get right into the mildew to see off the mould at the roots. Good old nature!

For this recipe make sure you purchase pure tea tree oil made from the Australian tree *Melaleuca alternifolia* – other types of tea tree are not powerful enough for the job. I've added lemon and rosemary to help balance out the strong scent of tea tree in this formula, but you can leave them out, if you prefer.

Put all the ingredients into the spray bottle and mix gently. Secure the pump lid.

To use

Shake well, then spray onto areas of mildew or mould and leave for 1 hour. Then, wipe down the affected surface with a damp cloth.

Using a cleaning spray of your choice (I prefer a vinegar-based spray, such as Kitchen Spray Your Way on page 28), rub the area with a clean, dry brush or cloth, or with a toothbrush if you need to get into the nooks and crannies. Keep going until the whole area is clean – it may take a bit of elbow grease, but the surfaces will come up beautifully.

* The best way to prevent mould and mildew growing in the first place is to have adequate ventilation. Try to get into the practice of opening your windows for at least 10 minutes every day. If you can, move furniture away from the walls so that air can circulate. In bath and shower areas, the best way to prevent mould is to dry off your tiles after each use.

GENERAL HOUSEHOLD **121**

Directory

Many ingredients can be picked up easily
from supermarkets, chemists, discount stores
or hardware shops but for more specialist
ingredients, or if you want to buy ingredients in
bulk, then I turn to the Internet. Here are some of
my favourite retailers:

G Baldwin & Co

www.baldwins.co.uk
171–173 Walworth Rd, London SE17 1RW, UK

Baldwins stock amber glass bottles in every size you could possibly need, and ship worldwide. They also stock spray nozzles and pump tops, as well as a wide range of essential oils and liquid castile soaps.

Summer Naturals

www.summernaturals.co.uk (online only)

Summer Naturals sell bulk supplies of most of the ingredients needed to make practically every cleaning product. Look out for bulk supplies of liquid castile soap (including slightly cheaper unbranded liquid castile soap), white vinegar, borax substitute, soda crystals, citric acid, bicarbonate of soda, witch hazel, essential oils and even beeswax pellets.

eBay

www.ebay.co.uk (online only)

eBay is a handy place to find really cheap bulk supplies of white vinegar, and you can also find bulk supplies of borax, soda crystals, citric acid, bicarbonate of soda, witch hazel, beeswax pellets and essential oils easily and cheaply.

Holland & Barrett

www.hollandandbarrett.com (stores nationwide)

On the high street I find Holland & Barrett particularly useful for stocking up on standard sizes of essential oils, especially when they run their frequent sales.

Index

Acknowledgements

Many thanks to the team at Pavilion, especially Stephanie and Bella, for helping me bring this book to life.

Huge thanks to my partner Dale, who didn't bat an eyelid when I told him all those years ago that I wanted to wash our clothes in vinegar. His enthusiasm and support has never waned, even through the ill-fated homemade dishwasher detergent experimentation era.

To my daughters E and A, for their endless enthusiasm and understanding and patience whilst writing this book.

To my parents, who have always encouraged and supported me, no matter what.

And finally, thanks to my dear friend and chemistry teacher extraordinaire, Scott McGarvey, for his invaluable chemistry advice and support.